United States
Department of
Agriculture

Forest Service

Pacific Southwest
Research Station

General Technical
Report
PSW-GTR-235
June 2011

A Review of the Potential Effects of Climate Change on Quaking Aspen (*Populus tremuloides*) in the Western United States and a New Tool for Surveying Aspen Decline

Toni Lyn Morelli and Susan C. Carr

Authors

Toni Lyn Morelli was a research ecologist, Pacific Southwest Research Station, 800 Buchanan St., Albany, CA 94710; **Susan C. Carr** is a research ecologist, Rocky Mountain Research Station, 240 West Prospect St., Fort Collins, CO 80526. Morelli currently is located at the University of California, Berkeley

Cover Photo

Cover picture taken in 2008 in Terror Creek, on the Paonia Ranger District of the Grand Mesa Uncompahgre Gunnison National Forest, by Wayne Shepperd, Rocky Mountain Research Station, USDA Forest Service.

Abstract

Morelli, Toni Lyn; Carr, Susan C. 2011. A review of the potential effects of climate change on quaking aspen (*Populus tremuloides*) in the Western United States and a new tool for surveying sudden aspen decline. Gen. Tech. Rep. PSW-GTR-235. Albany, CA: U.S. Department of Agriculture, Forest Service, Pacific Southwest Research Station. 31 p.

We conducted a literature review of the effects of climate on the distribution and growth of quaking aspen (*Populus tremuloides* Michx.) in the Western United States. Based on our review, we summarize models of historical climate determinants of contemporary aspen distribution. Most quantitative climate-based models linked aspen presence and growth to moisture availability and solar radiation. We describe research results pertaining to global climate change effects on aspen distribution and vigor. In addition, we present potential interactive effects related to climate change and natural disturbances and insect and pathogen outbreaks. Finally, we review the phenomenon of sudden aspen decline in western North America, which has been linked to drought and may be exacerbated by future climate change. Overall, research indicates a complex, unpredictable future for aspen in the West, where increased drought, ozone, and insect outbreaks will vie with carbon dioxide fertilization and warmer soils, resulting in unknown cumulative effects. Considering its positive moisture influence on the landscape, its economic impact, and its many benefits to the resilience of wildlife in terms of habitat and forage, aspen is a valuable, yet vulnerable, species in the face of global warming.

Keywords: Drought, forest health, global warming, Rocky Mountains, Sierra Nevada, sudden aspen decline.

Summary

Given aspen's desirable effects on the landscape in terms of maintaining biodiversity and conserving water, aspen stands may become increasingly valuable resources in the face of climate change. However, moisture stress and root damage resulting from a warmer climate may lead to the decline of aspen stands. These negative impacts of warmer and drier climates could be offset by the direct benefits of elevated carbon dioxide (CO_2) on aspen growth and more frequent fires, which may increase aspen extent. On the other hand, if migration and regeneration or seedling establishment rates are sufficient to adapt to environmental change, aspen distribution may simply shift in response to future climate. Overall, research indicates a complex, unpredictable future for aspen in the West, where increased drought, ozone, and insect outbreaks will vie with CO_2 fertilization and warmer soils, resulting in unknown cumulative effects.

Content

Introduction

Despite its wide range and economic and ecologic importance, little is known about the influence of climate on the growth and reproduction of quaking aspen (*Populus tremuloides* Michx.) in western North America. Studies related to climate effects on aspen distribution are typically models of correlative effects and are clustered in a few study regions (e.g., the central Rocky Mountains and Northwest Canada). Predicted effects of climate change on aspen are even more poorly understood. Here we summarize the results of climate effects studies related to aspen distribution and growth, with particular emphasis on the phenomenon of sudden aspen decline (SAD) and its connection with climate. Our literature review includes conclusions and predictions relative to climate change effects on aspen populations in western North America. We focus primarily on the relationship between climate determinants and aspen distribution, including the interaction of climate factors with disturbance effects.

Aspen is the most widespread tree species in North America (Little 1971, Mueggler 1988). It is clonal, reproducing by root sprouting (Schier et al. 1985). Aspen clones thrive in high-resource environments, specifically high light and nutrient levels (Kinney et al. 1997). As a result, two-thirds of western aspen stands are seral, giving way to conifers that gradually overtop and shade out aspen starting after about 80 years (Mueggler 1985, Rogers 2002). In spite of this, distinctive features like clonality and shallow, widespread root systems make aspen a highly resilient species, as stems destroyed by pathogens, insects, or fire are replaced by root sprouts (Lieffers et al. 2001). Thus, in the absence of conifer competition, approximately 30 percent of aspen stands in the Western United States are hypothesized to be stable instead of seral, persisting in the absence of disturbance or climate change (Kay 1997, Sawyer and Keeler-Wolf 1995). Studies in the Rocky Mountain region support the idea that aspen regeneration can occur independently of natural disturbance (Elliot and Baker 2004, Kurzel et al. 2007, Larsen and Ripple 2003, Turner et al. 2003).

Aspen forests can be managed for multiple uses because they are associated with water conservation, livestock and wildlife forage, aesthetic value, and economic benefits through increased tourism (Bartos and Campbell 1998, DeByle and Winokur 1985). Moreover, aspen stands have been shown to be hotspots of biodiversity (Stohlgren et al. 1999), with aspen considered a keystone species (Bartos 2001). Aspen forests have higher vascular plant species richness than other forest communities of the southern Boreal region (Reich et al. 2001) and support some of the highest diversity in the Sierra Nevada (Potter 1998).

Aspen forests can be managed for multiple uses because they are associated with water conservation, livestock forage, aesthetic value, economic benefits, and are hotspots of biodiversity.

Aspen stands provide wildlife habitat in the form of structural diversity, cavities for nesting, decay among live stems, and a dense understory. Over 100 vertebrate and invertebrate herbivore species can be found in aspen forests (Lindroth 2008). Aspen stands support some of the highest bird diversity in the United States (De-Byle and Winokur 1985, Griffis-Kyle and Beier 2003) and the greatest number of bird species in the Sierra Nevada specifically (Richardson and Heath 2004). A Colorado study showed that aspen habitat contains more plant and butterfly species per area than any of the other major vegetation types (Chong et al. 2001), although soil moisture has an effect (Weixelman et al. 1999). Aspen is also important beaver (*Castor canadensis*) habitat (Shepperd et al. 2006).

Aspen stands can help conserve water on the landscape as well. Net water consumption by aspen trees, in terms of ground water and surface waterflow, is considerably less than that of conifers (Jaynes 1978, LaMalfa and Ryel 2008). Researchers have reported decreases of 7.62 to 17.78 cm (Gifford et al. 1984) in water yield to the watershed when conifers replace aspen. Aspen's water conservation is mostly due to its low water efficiency and also to greater snow accumulation under aspen (LaMalfa and Ryel 2008); tower-based monitoring of Canadian boreal forest sites (Amiro et al. 2006) showed greater annual evapotranspiration from aspen forest than from coniferous forests (black spruce [*Picea mariana* (Mill.) Britton, Sterns & Poggenburg] and jack pine [*Pinus banksiana* Lamb.]). Because of their higher moisture content and associated herbaceous understory, aspen stands also act as fire breaks (Fechner and Barrows 1976, Peet 2000, van Wagner 1977); aspen-dominated landscapes are less likely to ignite from lightning fire than spruce-dominated landscapes (Krawchuk et al. 2006). In fact, aspen stands have been found to be 200 times less likely to burn than spruce-fir stands (Bigler et al. 2005). Finally, soil under aspen stands retains more nutrients, such as nitrogen, potassium, and calcium than soil under conifers (St. Clair 2008).

An ongoing debate is whether aspen populations in the Western United States are decreasing outside the range of recent natural variability. Many researchers have shown evidence of substantial declines in aspen extent since the mid-19th century (Bartos 2001, Gallant et al. 2003, Potter 1998, Wirth et al. 1996), although not in all areas (Brown et al. 2006, Manier and Laven 2001). Some researchers hypothesize that the discrepancy between current and late 19th-century aspen extent reflects not unusual decline in recent decades but merely a return to historical levels after an exceptionally large pulse of aspen regeneration from around 1850 to 1920 as a result of extensive logging, mining, grazing, and burning in the Western United States (Kulakowski et al. 2004, 2006; Shepperd et al. 2006; Smith and Smith 2005).

Another factor in aspen distributon shifts may have been climate change. Cooler and wetter climate conditions at the end of the Little Ice Age appear to have increased aspen extent in the Sierra Nevada (Shepperd et al. 2006). The climate trends over the last several decades of increasing temperature and reduced moisture may explain some of the more recent decline in aspen extent (Hogg et al. 2008, Worrall et al. 2008). Although the different hypotheses of human disturbance and changing climate are difficult to separate (Millar and Woolfenden 1999), photographic and other evidence confirms that aspen were generally more prevalent throughout the Sierra Nevada in the recent past than they are now (Shepperd et al. 2006), similar to other areas of the West.

Climate Mitigates Aspen Extent

General environmental conditions: aspen growth and distribution in western North America—

Climate is a strong determining factor for the growth and distribution of aspen in western North America. In general, long-term trends in temperature, precipitation, and solar radiation, coupled with environmental conditions, affect the availability of aspen habitat. Although aspen can tolerate extremely cold air temperatures, cold soils (6 °C or less) stress aspen plants, leading to inhibited root growth and decreased water intake (Landhäusser and Lieffers 1998, Wan and Zwiazek 1999). Moreover, despite the size and motion of the leaves that prevent overheating and stomatal closure and give aspen its name (Roden and Pearcy 1993), aspen trees function poorly in hot, dry conditions (Jones et al. 1985b). Photosynthesis declines at temperatures greater than 25 °C (Lawrence and Oechel 1983), especially when humidity is low (Dang et al. 1997).

Aspen trees in western North America typically inhabit areas where annual precipitation exceeds annual evapotranspiration (Jones 1985). In general, stands in the Rocky Mountain region occur where total annual precipitation exceeds 38 cm per year (Jones 1985, Jones and DeByle 1985). Similarly, aspen stands in Canada occur where total precipitation exceeds potential evapotranspiration (PET, a climate moisture index) (Chen et al. 2002, Hogg 1994, Hogg and Hurdle 1995). Aspen distribution is also related to growing-season precipitation and moisture deficit in the Great Lakes region of the United States and Canada (Gustafson et al. 2003, Iverson and Prasad 1998).

The influence of topography and landscape orientation on aspen distribution further underscores the effects of temperature and moisture availability. Elevation, aspect, and slope affect local climate environment, including length of the frost-free period and temperature extremes. Aspen is restricted to higher elevations and more northerly aspects in the warmer southern regions of its distribution (i.e., the southern Rocky Mountains and Baja California), whereas aspen stands commonly inhabit south-facing slopes at higher elevations and in the colder parts of their extent (Jones et al. 1985b). For example, in the Rocky Mountains, aspen stands typically occupy elevations between 1828 and 3048 m (Jones 1985), whereas in the boreal and prairie transition regions of northwest Canada, aspen inhabits lower elevations (less than 1828 m) (Perala 1990).

Aspen distribution appears related to local edaphic conditions as well as climate. In western North America, aspen stands inhabit soils that are well-drained, loamy, high in organic matter, and have soil water tables between 0.6 and 2.5 m (Perala 1990). In the central and northern Rocky Mountain region, aspen stands occur on soils derived from basic igneous rock or neutral to calcareous shales and limestones. However, aspen stands also inhabit riparian and other poorly drained areas, likely owing to the consistent water supply as well as the lack of conifer competition there.

Regional ecological models of aspen distribution and growth—
Ecological models from specific regions in the West underscore the influence of climate on aspen extent (table 1). Two regions where aspen-climate relationships have been well-studied are the northern Rocky Mountains and the boreal and prairie regions of western Canada. The resulting correlative models differ with respect to which climate variables were considered and which climate variables were found to be related to aspen presence and growth. However, in general, studies show that aspen distribution appears to be related to temperature, precipitation, and solar radiation. In the Greater Yellowstone Ecosystem (GYE) of the northern Rocky Mountains, growing season shortwave radiation was the primary factor correlated with aspen presence, coinciding with a major north-south split in aspen distribution and abundance (Brown 2003, Brown et al. 2006). Brown (2003) found that GYE aspen occurs in warmer and wetter sites compared to coniferous species, with higher amounts of solar radiation, snowfall, and temperatures, and lower values of PET. Specifically, GYE aspen stands occurred at an average elevation of 2300 m (range = 1559 to 2921 m), with an annual precipitation of 70.6 cm (range = 33.8 to 153.4 cm) and mean annual temperature of 2.1 °C (range = 2.2 to 6.1 °C). Similarly, GYE aspen growth rates were positively correlated with temperature (annual maximum temperature 7 to 12 °C) (Brown 2003, Brown et al. 2006). The authors suggested that earlier onset of aspen spring growth in the GYE is associated with higher spring temperatures and precipitation.

4

Table 1—Publications that include numerical models of aspen performance related to climate predictors[a]

Reference	Study region	Model/response variable(s)	Significant climate predictors	Comments
Brown 2003, Brown et al. 2006	GYE	Model 1: Classification tree regression of aspen presence/absence	GS shortwave radiation, GS PET, annual snowfall, GS min temperature, slope	GS shortwave radiation correlated with geographic separation (North and South). GS PET most important in southern stands
	GYE	Model 2: Multiple regression of aspen growth (determined from tree ring cores)	Annual max temperature, GS PET, slope	Highest aspen growth on sites with warmer temperatures and intermediate GS precipitation
	GYE	Model 3: Multiple regression of percentage of aspen cover change between 1956 and 2001	Total annual snowfall, percentage of conifer change, GS shortwave radiation	Aspen cover decrease related to increased snowfall and conifer cover, aspen cover increase positively correlated with GS shortwave radiation
Brandt et al. 2003	Canada Prairie region	Multiple regression of climate and pest predictors on percentage of dead and living stems	CMI, years pest infestation (root disease and tent caterpillar), tree age	Percentage of dead stems negatively correlated with CMI, percentage of living stems positively correlated with CMI
Elliott and Baker 2004	SW Colorado; treeline in San Juan Mountains	Multiple subset regression of aspen: year of seedling establishment	Mean spring, summer, and annual precipitation	Aspen seedling establishment occurred in cooler years with higher spring precipitation; vegetative growth in drier, warmer years
Gustafson et al. 2003	Upper Great Lakes	Multiple regression of site and climate predictors over a large region: annual growth estimates from tree ring analysis	Topographic moisture index, GS precipitation, number of optimum growing days, June moisture deficit, soil drainage class	Statistical model includes predictor variables of aspen growth

Table 1—Publications that include numerical models of aspen performance related to climate predictors[a] (continued)

Reference	Study region	Model/response variable(s)	Significant climate predictors	Comments
Hessl and Graumlich 2002	GYE	Chi-square tests of predictor variables across aspen age classes estimated from tree cores	None	Only climate predictors considered were total precipitation and Palmer drought stress index. Statistical tests and predictor variables very general
Hogg et al. 2005	Western Canada	Multiple regression of annual aspen growth estimated from tree ring analysis	Current year CMI, CMI from previous 4 years, canopy defoliation, growing season degree days April–July	CMI was the best predictor of aspen growth over a 50-year period
Hogg et al. 2008	Western Canada	Model 1: Multiple regression of climate and defoliation predictors on aspen growth (net and total biomass change, 2000–2005)	Minimum annual climate moisture index 2000–2004, percentage of canopy defoliation, mean stand age, percentage of silt	Aspen growth and mortality during a 5-year period was most affected by drought
		Model 2: Multiple regression of climate and defoliation percentage of stem mortality, 2000–2005		
Hogg 1994, Hogg and Hurdle 1995	Western Canada Parkland region	Overlay of climate isoclines with current aspen range; prediction of future isoclines and corresponding aspen distribution	CMI	Current aspen range coincident with CMI > -15 cm (-5.9 in) isocline; predicted northern shift in CMI isoclines under 2 x carbon dioxide (CO_2) model, reduction of boreal (aspen) forest by one half, resembling contemporary parkland region

Table 1—Publications that include numerical models of aspen performance related to climate predictors[a] (continued)

Reference	Study region	Model/response variable(s)	Significant climate predictors	Comments
Iverson and Prasad 1998	Eastern United States	Regression tree model of aspen importance values derived from Forest Inventory and Analysis data by county; predictive model of future distribution based on same model	Mean annual temperature, heterogeneity of county elevation, mean annual PET, soil texture	Mean annual temperture most influential at coarse scale (< 4.4 °C/7.9 °F), followed by flat topography, PET < 60 mm/mo (<2.4 in/mo) and prevalence of sandy soils. Prediction of extreme range reduction around Great Lakes region under 2 x CO_2 climate change
Sexton et al. 2006	Eastern Utah	Geographic information system model of predicted aspen presence as function of PET and conifer cover	September PET most discriminating variable of aspen vs. conifer presence	N/A

Note: Some publications present more than one model (listed singularly by row).

P = precipitation, PET = potential evapotranspiration, GS = growing season, GYE = Greater Yellowstone Ecosystem, CMI = climate moisture index, equal to the monthly precipitation minus PET.

[a] Model types and response variables are summarized, as well as climate variables found to be significant in the model.

A recent indepth analysis (Rehfeldt et al. 2009) of aspen climate space tested a 34-variable model to identify the predictors of current aspen extent in the U.S. Forest Service's Rocky Mountain Region, based on 118,000 U.S. Forest Inventory and Analysis (FIA) plots. A subset model that included eight climate variables successfully predicted most of the current extent of aspen. The annual dryness index (a ratio of growing-degree-days to annual precipitation) was the strongest climate predictor. Their general conclusion was that aspen distribution limits at the xeric fringe of its range are dictated mostly by moisture stress (Rehfeldt et al. 2009).

Ecological models in other regions suggest the influence of moisture and temperature on aspen distribution as well. Aspen stands in Canada exist where moisture availability is not limiting, as measured by total annual precipitation exceeding annual PET (Chen et al. 2002, Hogg 1994, Hogg and Hurdle 1995). Similarly, aspen distribution is related to growing-season precipitation and moisture deficit in the Eastern United States and Great Lakes region (Gustafson et al. 2003, Iverson and Prasad 1998).

Moisture availability appears to affect western aspen growth as well as distribution. Models of aspen growth are typically based on retrospective analysis of tree ring patterns, which is correlated with historical climate patterns (Brown 2003, Brown et al. 2006, Gustafson et al. 2003, Hessl and Graumlich 2002), and suggest that extreme drought conditions impose the greatest limits to aspen growth and survival (Brandt et al. 2003, Hogg et al. 2005, 2008). Studies of net ecosystem production in aspen support these findings (Barr et al. 2007). Climate moisture indices from the period of a recent severe drought in western Canada explained the most variation in recent aspen growth and mortality (Hogg et al. 2008). A study in Manitoba, Canada, found that a hot June reduced radial growth of quaking aspen, and aspen trees do not depend on early season water availability, for growth to the same extent as bur oak. The study also found that the temperature in the previous October does not necessarily affect aspen tree growth (*Quercus macrocarpa* Michx.) (Boone et al. 2004).

How Will Aspen Respond to Future Climates?

There are some common expectations for how the climate of western North America will change as atmospheric carbon dioxide (CO_2) increases. Overall, climatologists project that the Western United States will see increased summer temperatures, more precipitation in the form of rain and less snow, lower total annual precipitation in most areas, and increasing extreme weather prompting more frequent natural disturbances (Cayan et al. 2008, Dettinger 2005, Knowles and Cayan 2004, Mastrandrea et al. 2009, Moser et al. 2009). Here we summarize predictions for aspen stand dynamics for the next century in light of generalized climate projections.

Temperature and precipitation—
Higher temperatures and increased moisture stress are predicted to affect aspen mortality and regeneration in western North America (Brandt et al. 2003, Elliott and Baker 2004, Worrall et al. 2008). Aspen is a water-limited, drought-intolerant species (Niinemets and Valladares 2006); thus, severe droughts can cause death or decline of aspen. Such drought impacts have been seen in Canada (Hogg et al. 2002, Zoltai et al. 1991), where increased temperatures and changes in precipitation patterns coincided with reduced aspen presence in Canadian boreal forests (Hogg 1994, Hogg and Hurdle 1995). Increased evapotranspiration and decreased moisture have been implicated in the conversion of Canadian aspen parklands to grassland (Zoltai et al. 1991). Decreased moisture availability is predicted to disfavor aspen in eastern Utah, because of its higher water demands compared to co-occurring conifer species (Sexton et al. 2006). Similar evidence from the Eastern United

Climatologists project that the Western United States will see increased summer temperatures, more precipitation in the form of rain and less snow, lower total annual precipitation in most areas, and increasing extreme weather prompting more frequent natural disturbances.

States has prompted predictions of the eventual disappearance of aspen from that region (Iverson et al. 2008a, 2008b). A modeling analysis in Wisconsin predicted that warming climates could cause aspen to decline in the boreal forest of the Great Lakes region (He et al. 2002).

Changes in winter precipitation may have negative impacts on aspen as well. Snow cover mediates soil temperature, providing insulation for roots in extreme cold (Frey et al. 2004), and inhibiting ungulate browsing in winter (Martin 2007). Thus, reduced snow accumulation may contribute to damaged roots and retard regeneration. However, the relationships are unclear: one analysis of aspen extent showed that mild winters and warmer wetter summers favored aspen, and snowy cold winters and dry bright summers were detrimental to aspen, leading to grasslands and conifer succession (Brown et al. 2006).

Increased atmospheric CO$_2$—

As CO$_2$ increases, longer roots and thus better nutrient uptake should increase aspen productivity (Pregitzer et al. 2000). One experimental study in Wisconsin showed that aspen growth increased 39 percent with elevated atmospheric CO$_2$, with a faster rate under increased moisture (Norby et al. 2005). However, benefits may decrease over time, and increased ozone may negate the positive effects of elevated CO$_2$ (Kubiske et al. 2006). One study modeled that aspen in the Canadian boreal will increase productivity for the next 200 years, acting as a large carbon sink. However, prolonged (6-year) droughts would eventually cause severe dieback (Grant et al. 2006). Therefore, some researchers stress that the long-term effects of elevated atmospheric CO$_2$ on aspen will be complex and difficult to predict (Hogg 2001, Lindroth et al. 2001).

Climate interactions with natural disturbances—

Future climate changes may increase the frequency of physical disturbances (e.g., floods and wildfire), which alone would be expected to increase aspen on the landscape. However, interactions between different factors make the net effect of extreme weather difficult to predict. If the climate warms and dries, and if there are other stressors present such as heavy ungulate browsing, aspen may be unable to resprout or establish new seedlings (Romme et al. 2001).

Changes in fire frequency are predicted to affect aspen distribution and growth. Many authors have argued that increased temperatures and decreased precipitation would lead to more frequent fires (Spracklen et al. 2009, Westerling et al. 2006), which would favor aspen regeneration through suckering (Elliot and Baker 2004, Graham et al. 1990, Jones and DeByle 1985, Rogers 2002, Schier et al. 1985). In fact, in high-elevation forests with long fire intervals, the natural succession of aspen stands to conifers may be reset by future stand-replacing fires, especially if the area burned by such fires increases because of climate change (Dale et al. 2001).

Insects and other interaction effects—
Climate change may induce indirect effects on aspen productivity via increased frequency of and vulnerability to pathogens and herbivores, which interact with environmental stress to cause tree mortality (Frey et al. 2004; Hogg et al. 2005, 2008; Jones et al. 1985a). For instance, drought conditions in the spring and following summer or deep late spring snowpacks plus summer drought may increase the susceptibility of aspen to death through canker infections (Cryer and Murray 1992, Johnston 2001). Moreover, drier, warmer conditions may favor gypsy moth (*Lymantrai dispar*) invasions in Utah and possibly elsewhere in the West (Logan 2008) and forest tent caterpillar (*Malacosoma disstria* Hubner) outbreaks in western Canada (Hogg et al. 2002). Drought could also reduce sprouting after a disturbance because of higher susceptibility to insects and pathogens (Sexton et al. 2006).

Mammal herbivory can exacerbate drought effects on aspen growth and distribution. For example, chronic heavy browsing by elk (*Cervus canadensis*) in the interior Western United States (e.g., Rocky Mountain National Park), in combination with drought and fire suppression, seems to be leading to aspen decline (Romme et al. 2001). Climate change may have the strongest effect on areas where aspen are patchily distributed on marginal habitat and ungulate browsing is heavy (Romme et al. 2001). The Book Cliffs in Utah exemplify the combination of drying climates, displacement by conifers through shading, soil and microclimate effects, and ungulate browsing that could decrease aspen cover in the future (Sexton et al. 2006).

The role of seedling regeneration—
Aspen seedling regeneration may become increasingly important in a changing climate, providing the genetic diversity needed for the population to adapt to rapidly changing conditions. There is debate over the average age of aspen clones, but if current aspen stands were established in centuries past, they may be genetically adapted to cooler climates such as occurred during the Little Ice Age (Barnes 1966, Tuskan et al. 1996). A rare quantitative analysis showed that aspen seedling establishment at treeline in southwestern Colorado may have occurred in years with lower than normal mean maximum summer temperatures (21 to 22 °C) and higher mean spring precipitation (5 to 6 cm). Conversely, accelerated asexual reproduction was correlated with lower spring precipitation (3 to 4 cm) and warmer maximum summer temperatures (23 to 24 °C) (Elliott and Baker 2004). The researchers

speculated that future aspen seedling regeneration may be limited to higher elevations and latitudes where the requisite cooler and wetter temperatures prevail.

Sudden Aspen Decline (SAD)

An ongoing phenomenon, the rapid death of some or all of a mature aspen stand with little or no regeneration, dubbed sudden aspen decline (SAD), may be an indicator of the response of aspen to climate change. It was brought into focus in Utah and Arizona starting in 2002, and soon after in Colorado (Shepperd 2008, Worrall et al. 2008). However, unusual aspen mortality has occurred periodically over the last four decades in the Great Lakes region, Canada, and the interior Western United States.

An ongoing phenomenon, the rapid death of some or all of a mature aspen stand with little or no regeneration, dubbed sudden aspen decline (SAD), may be an indicator of the response of aspen to climate change.

Characteristics—

Sudden aspen decline occurs rapidly and simultaneously across a grove, in 1 to 3 years (Peterson and Peterson 1992, Worrall et al. 2008). It appears on the landscape as white defoliated trees that are still standing with their bark intact, indicating that they died recently. Large trees appear to die first and effects may start at the edge of a grove (Ciesla 2008). Younger cohorts are often not affected (Shepperd and Guyon 2006).

One can distinguish SAD from insect defoliation or frost damage because of complete defoliation in addition to dieback of tree branches (Worrall et al. 2008). There is great concern among researchers that roots are dying first (Worrall et al. 2008), resulting in the lack of regeneration and other stereotypical SAD signs, although the response may be delayed for a season (Campbell et al. 2008). A 2007 study found up to 90 percent of root volume dead in several stands in Colorado (Worrall et al. 2008). With complete root death, the aspen grove will eventually revert to a nonaspen vegetation type.

Incidence—

Sudden aspen decline has occurred recently and most noticeably in southwestern Colorado, northern Arizona, and parts of Utah and Canada, but it has also been seen in Idaho, Nevada, Montana, and Wyoming. Data from aerial detection surveys of permanent plots indicated that the average mortality rate of aspen in Utah, Nevada, and western Wyoming in 2006 and 2007 was 31 percent; two-thirds of all dead trees died between 2005 and 2007 (Hoffman et al. 2008). A 2006 aerial survey across Colorado spotted 56 091 ha of SAD (Worrall et al. 2008). An estimated 13 percent of aspen cover in Colorado showed effects of SAD by 2007 (Rodebaugh 2008), and a 2008 aerial survey revealed that 216 000 ha were noticeably affected (http://www.aspensite.org/SAD/sad_faqs.pdf).

Outside of the intermountain West and the Rocky Mountains, the extent of SAD is unclear. A recent survey in eastern Washington showed no sign of SAD in two national forests (Hadfield and Magelssen 2004). Very little survey work has been done to explore the incidence of SAD in California. To aid survey efforts, we have developed a new SAD survey tool to help federal employees obtain baseline data from aspen stands and identify SAD events as they emerge (see appendix).

Causes—

Sudden aspen decline appears to have a strong climate correlation, as most occurrences can be related to high temperatures and drought (Worrall et al. 2008). In addition, SAD-like events seem to occur earlier in areas with higher annual temperatures and drier climates (Hogg and Hurdle 1995, Shields and Bockheim 1981). One explanation for SAD in the intermountain West is that drought and hot weather in the early 2000s stressed aspen stands. A related cascade of events was seen at the same time in western Canada (Hogg et al. 2008). Similarly, a drought in 1961 caused ubiquitous aspen mortality in the grasslands of western Canada a few years later, causing direct deaths or secondary deaths from *Cytospora* canker (Zoltai et al. 1991). A study by Rehfeldt et al. (2009), using several general circulation models and climate scenarios, suggested that most SAD events in the Rocky Mountain region occurred within areas that they project may no longer be viable aspen habitat by 2060.

Researchers have developed a decline disease hypothesis for SAD (Frey et al. 2004, Worrall et al. 2008): stand and site factors such as age, slope, and aspect predispose aspen to decline; defoliation or a severe drought and high summer temperatures incite decline among those aspen predisposed; and finally, opportunistic insects and pathogens contribute to the death of the aspen. Research has implicated other factors, including herbivore impacts and freeze-thaw events (Cayford et al. 1959, Cox and Malcolm 1997, Frey et al. 2004). Fine root damage caused by extreme winter freeze followed by drought could cause SAD by reducing water and nutrient uptake (Frey et al. 2004). A comparable winter exposure mechanism has been implicated in the sudden decline of yellow-cedar (*Chamaecyparis nootkatensis* (D. Don) Spach) in southeastern Alaska (Beier et al. 2008).

Some stands and sites in the Western United States are particularly vulnerable to SAD (Baker and Shaw 2008, Brandt et al. 2003, Worrall et al. 2008): those at (1) low elevation (e.g., 2100 to 2500 m in the Colorado Rocky Mountains), (2) south and southwest aspects, and (3) flatter slopes. Over 90 percent of aspen stems have died on some low-elevation sites in Arizona, with 16 to 43 percent mortality in mid- and high-elevation sites (Fairweather and Geils 2008). Because low-elevation sites and southern aspects are generally drier and warmer in the summer and more prone to additional stress by freezing and drying soil in the winter, these observations suggest a climate causation. Aspen stands on sloped areas may be better adapted to moisture stress and thus not as affected by acute drought events (Worrall et al. 2008). Further, with changing climates low-elevation sites may be receiving less snow and thus may be increasingly less insulated and more vulnerable to freeze-thaw events during the winter. There may also be a correlation with conifer competition, as conifers may not be as abundant at high elevations.

Although there has been some debate (Frey et al. 2004), results indicate that SAD vulnerability does not increase with age once trees are physiologically mature (Brandt et al. 2003, Worrall et al. 2008). There is further uncertainty on whether or not tree size is correlated with SAD. Some researchers have shown large-diameter trees (stems greater than 30 cm) to be more susceptible (Worrall et al. 2008), whereas others hypothesize that tall thin trees in exposed xeric sites would be most vulnerable (Frey et al. 2004). If drought is the inciting factor for SAD, trees with small diameter at breast height should be most affected because water stress would increase water tension and xylem cavitation and cause dieback in the upper crown first, e.g., in cottonwoods (*Populus* spp.) (Rood et al. 2000). Further, conflicting results point to a potential effect of stand density (Hogg et al. 2002, Worrall et al. 2008), although there is a potential confounding of latitude and insect preferences.

Various pathogens and insects appear to be more commonly associated with SAD than with other aspen mortality (Worrall et al. 2008). Although no single biotic factor appears to be responsible for SAD, five organisms were found to be commonly associated with SAD in Colorado (Worrall et al. 2008): *Cytospora* canker, poplar borer (*Saperda calcarata*), bronze poplar borer (*Agrilus liragus*), and two aspen bark beetle species (*Trypophloeus populi* and *Procryphalus mucronatus*). All five are species that do not normally cause mortality in healthy aspen.

Conclusion

Climate change, through increased drought, ozone, and insect outbreaks, may cause aspen to become increasingly threatened (Nitschke and Innes 2008), exemplified by the SAD phenomenon. Conversely, elevated CO_2 and more frequent fires could increase aspen extent (Shepperd et al. 2006, Zoltai et al. 1991). Alternatively, aspen distribution may simply shift in the future (Rehfeldt et al. 2009, Ryel and Bartos 2008) if migration and regeneration or seedling establishment rates are sufficient to adapt to environmental change (Iverson et al. 2004). Changes in suitable aspen habitat will likely differ by region (Hogg 2001), necessitating decentralized approaches to research, monitoring, and management. Our review underscores the usefulness of local knowledge regarding aspen management, particularly in regard to predicting where aspen may thrive in the future. Finally, the survey we present here can help land managers track the SAD phenomenon throughout western North America, especially if SAD increases as a threat in a warmer, drier future.

Acknowledgments

This review was conducted as part of a larger study of climate change effects on aspen distribution, sponsored by the USDA Forest Service, Rocky Mountain and Pacific Northwest Research Stations. Susan Frankel, Edward (Ted) Hogg, Patricia Manley, Michael Michaelian, and Wayne Shepperd provided extremely useful and timely comments on the manuscript. We thank Linda Joyce and Connie Millar for their contributions to this work. Morelli is grateful to John Guyon, Ted Hogg, Connie Millar, Paul Rogers, Wayne Shepperd, Jim Worrall, and especially David Burton for their contributions to the SAD survey.

Metric Equivalents

When you know:	Multiply by:	To get:
Millimeters (mm)	0.0394	Inches
Centimeters (cm)	0.394	Inches
Meters (m)	3.28	Feet
Hectares (ha)	2.47	Acres
Degrees Celsius (°C)	1.8C + 32	Degrees Fahrenheit

References

Amiro, B.D.; Orchansky, A.L.; Barr, A.G.; Black, T.A.; Chambers, S.D.; Chapin, F.S., III; Goulden, M.L.; Litvak, M.; Liu, H.P.; McCaughey, J.H.; McMillan, A.; Randerson, J.T. 2006. The effect of post-fire stand age on the boreal forest energy balance. Agricultural and Forest Meteorology. 140(1-4): 41–50.

Baker, F.A.; Shaw, J.D. 2008. Characteristics of aspen and aspen mortality [Poster]. In: Sudden aspen decline (SAD) meeting. Fort Collins, CO. http://www.aspensite.org/. (15 March 2010).

Barnes, B.V. 1966. The clonal growth habit of American aspens. Ecology. 47: 439–447.

Barr, A.G.; Black, T.A.; Hogg, E.H.; Griffis, T.J.; Morgenstern, K.; Kljun, N.; Theede, A.; Nesic, Z. 2007. Climatic controls on the carbon and water balances of a boreal aspen forest 1994–2003. Global Change Biology. 13: 561–576.

Bartos, D.L. 2001. Landscape dynamics of aspen and conifer forests. In: Shepperd, W.D.; Binkley, D.; Bartos, D.L.; Stohlgren, T.J.; Eskew, L.G., comps. Sustaining aspen in western landscapes: symposium proceedings. Gen. Tech. Rep. RMRS-P-18. Fort Collins, CO: U.S. Department of Agriculture, Forest Service, Rocky Mountain Research Station: 5–14.

Bartos, D.L.; Campbell, R.B., Jr. 1998. Decline of quaking aspen in the interior West—examples from Utah. Rangelands. 20(1): 17–24.

Beier, C.M.; Sink, S.E.; Hennon, P.E.; D'Amore, D.V.; Juday, G.P. 2008. Twentieth-century warming and the dendroclimatology of declining yellow-cedar forests in southeastern Alaska. Canadian Journal of Forest Research. 38(6): 1319–1334.

Bigler, C.; Kulakowski, D.; Veblen, T.T. 2005. Multiple disturbance interactions and drought influence fire severity in Rocky Mountain subalpine forests. Ecology. 86(11): 3018–3029.

Boone, R.; Tardif, J.; Westwood, R. 2004. Radial growth of oak and aspen near a coal-fired station, Manitoba, Canada. Tree-Ring Research. 60(1): 45–58.

Brandt, J.P.; Cerezke, H.F.; Mallet, K.I.; Volney, W.J.A.; Weber, J.D. 2003. Factors affecting trembling aspen (*Populus tremuloides* Michx.) health in Alberta, Saskatchewan, and Manitoba, Canada. Forest Ecology and Management. 178: 287–300.

Brown, K. 2003. Understanding the role of biophysical setting in aspen persistence. Bozeman, MT: Montana State University. 96 p. M.S. thesis.

Brown, K.; Hansen, A.J.; Keane, R.K.; Graumlich, L.J. 2006. Complex interactions shaping aspen dynamics in the greater Yellowstone ecosystem. Landscape Ecology. 21: 933–951.

Campbell, R.B., Jr.; Bartos, D.L.; Henningson, A.V. 2008. Aspen condition and management activities in southern Utah [Poster]. In: Sudden aspen decline (SAD) meeting. Fort Collins, CO. http://www.aspensite.org/. (15 March 2010).

Cayan, D.R.; Maurer, E.P.; Dettinger, M.D.; Tyree, M.; Hayhoe, K. 2008. Climate change scenarios for the California region. Climatic Change. 87(Suppl. 1): S21–S42.

Cayford, J.H.; Hildahl, V.; Nairn, L.D.; Wheaton, M.P.H. 1959. Injury to trees from winter drying and frost in Manitoba and Saskatchewan in 1958. The Forestry Chronicle. 35: 282–290.

Chen, H.Y.H.; Krestov, P.V.; Klinka, K. 2002. Trembling aspen site index in relation to environmental measures of site quality at two spatial scales. Canadian Journal of Forestry. 32: 112–119.

Chong, G.W.; Simonson, S.E.; Stohlgren, T.J.; Kalkhan, M.A. 2001. Biodiversity: Aspen stands have the lead, but will non-native species take over? In: Shepperd, W.D.; Binkley, D.; Bartos, D.L.; Stohlgren, T.J.; Eskew, L.G., comps. Sustaining aspen in western landscapes: symposium proceedings. Gen. Tech. Rep. RMRS-P-18. Fort Collins, CO: U.S. Department of Agriculture, Forest Service, Rocky Mountain Research Station: 261–272.

Ciesla, W.M. 2008. Aspen decline and other factors affecting the health of aspen on the Colorado Front Range [Poster]. In: Sudden aspen decline (SAD) meeting. Fort Collins, CO. http://www.aspensite.org/. (15 March 2010).

Cox, R.M.; Malcolm, J.W. 1997. Effects of duration of a simulated winter thaw on dieback and xylem conductivity of *Betula papyrifera*. Tree Physiology. 17: 389–396.

Cryer, D.H.; Murray, J.E. 1992. Aspen regeneration and soils. Rangelands. 14(4): 223–226.

Dale, V.H.; Joyce, L.A.; McNulty, S.; Neilson, R.P.; Ayres, M.P.; Flannigan, M.D.; Hanson, P.J.; Irland, L.C.; Lugo, A.E.; Peterson, C.J.; Simberloff, D.; Swanson, F.J.; Stocks, B.J.; Wotton, B.M. 2001. Climate change and forest disturbances. BioScience. 51: 723–734.

Dang, Q.L.; Margolis, H.; Coyea, M.R.; Sy, M.; Collatz, G.J. 1997. Regulation of branch-level gas exchange of boreal forest trees: roles of shoot water potential and vapor pressure differences. Tree Physiology. 17: 521–535.

DeByle, N.V.; Winokur, R.P. 1985. Introduction. In: DeByle, N.V.; Winokur, R.P., eds. Aspen: ecology and management in the Western United States. Gen. Tech. Rep. RM-119. Fort Collins, CO: U.S. Department of Agriculture, Forest Service, Rocky Mountain Forest and Range Experiment Station: 1.

Dettinger, M.D. 2005. From climate-change spaghetti to climate-change distributions for 21st century California. San Francisco Estuary and Watershed Science. 3(1): Article 4. Available at http://repositories.cdlib.org/jmie/sfews/vol3/iss1/art4: (15 March 2010).

Elliott, G.P.; Baker, W.L. 2004. Quaking aspen (*Populus tremuloides* Michx.) at treeline: a century of change in the San Juan Mountains, Colorado, USA. Journal of Biogeography. 31: 733–745.

Fairweather, M.L.; Geils, B. 2008. Aspen decline in northern Arizona [Poster]. In: Sudden aspen decline (SAD) meeting. Fort Collins, CO. http://www.aspensite.org/. (15 March 2010).

Fechner, G.H.; Barrows, J.S. 1976. Aspen stands as wildfire fuel breaks. Eisenhower Consortium Bulletin 4. Resour. Bull. Fort Collins, CO: U.S. Department of Agriculture, Forest Service, Rocky Mountain Forest and Range Experiment Station. 29 p.

Frey, B.R.; Leiffers, V.J.; Hogg, E.H.; Landhausser, S.M. 2004. Predicting landscape patterns of aspen dieback: mechanisms and knowledge gaps. Canadian Journal of Forestry. 34: 1379–1390.

Gallant, A.L.; Hansen, A.J.; Councilman, J.S.; Monte, D.K.; Betz, D.W. 2003. Vegetation dynamics under fire exclusion and logging in a Rocky Mountain watershed, 1856–1996. Ecological Applications. 13(2): 385–403.

Gifford, G.F.; Humphries, W.; Jaynes, R.A. 1984. A preliminary quantification of the impacts of aspen to conifer succession on water yield. II. Modeling results. Water Resources Bulletin. 20: 181–186.

Graham, R.L.; Turner, M.G.; Dale, V.H. 1990. How increasing atmospheric CO_2 and climate change affect forests. BioScience. 40: 575–587.

Grant, R.F.; Black, T.A.; Gaumont-Guay, D.; Klujn, N.; Barr, A.G.; Morgenstern, K.; Nesic, Z. 2006. Net ecosystem productivity of boreal aspen forests under drought and climate change: mathematical modeling with Ecosys. Agricultural and Forest Meteorology. 140: 152–170.

Griffis-Kyle, K.L.; Beier, P. 2003. Small isolated aspen stands enrich bird communities in southwestern ponderosa pine forests. Biological Conservation. 110(3): 375–385.

Gustafson, E.J.; Lietz, S.M.; Wright, J.L. 2003. Predicting the spatial distribution of aspen growth potential in the upper Great Lakes region. Forest Science. 49(4): 499–508.

Hadfield, J.; Magelssen, R. 2004. Assessment of aspen condition on the Okanogan and Wenatchee National Forests. Wenatchee, WA: U.S. Department of Agriculture, Forest Service, Okanogan and Wenatchee National Forests. 27 p.

He, H.S.; Mladenoff, D.J.; Gustafson, E.J. 2002. Study of landscape change under forest harvesting and climate warming-induced fire disturbance. Forest Ecology and Management. 155: 257–270.

Hessl, A.E.; Graumlich, L.J. 2002. Interactive effects of human activities, herbivory and fire on quaking aspen (*Populus tremuloides*) age structures in western Wyoming. Journal of Biogeography. 29: 889–902.

Hoffman, J.T.; Guyon, J.C.; Steed, B. 2008. Monitoring the condition of aspen in the Northern and Intermountain Regions [Poster]. In: Sudden aspen decline (SAD) meeting. Fort Collins, CO. http://www.aspensite.org/. (15 March 2010).

Hogg, E.H. 1994. Climate and the southern limit of the western Canadian boreal forest. Canadian Journal of Forest Research. 24(9): 1835–1845.

Hogg, E.H. 2001. Modeling aspen responses to climatic warming and insect defoliation in western Canada. In: Shepperd, W.D.; Binkley, D.; Bartos, D.L.; Stohlgren, T.J.; Eskew, L.G., comps. Sustaining aspen in western landscapes: symposium proceedings. Gen. Tech. Rep. RMRS-P-18. Fort Collins, CO: U.S. Department of Agriculture, Forest Service, Rocky Mountain Research Station. 460 p.

Hogg, E.H.; Brandt, J.P.; Kochtubajda, B. 2002. Growth and dieback of aspen forests in northwestern Alberta, Canada, in relation to climate and insects. Canadian Journal of Forest Research. 32: 823–832.

Hogg, E.H.; Brandt, J.P.; Michaelian, M. 2005. Factors affecting inter-annual variation in growth of western Canadian aspen forests during 1951–2000. Canadian Journal of Forestry. 35: 610–622.

Hogg, E.H.; Brandt, J.P.; Michaelian, M. 2008. Impacts of a regional drought on productivity, dieback, and biomass of western Canadian aspen forests. Canadian Journal of Forestry. 38: 1373–1384.

Hogg, E.H.; Hurdle, P.A. 1995. The aspen parkland in western Canada: a dry-climate analogue for the future boreal forest? Water, Air, and Soil Pollution. 82: 391–400.

Iverson, L.R.; Prasad, A.M. 1998. Predicting abundance of 80 tree species following climate change in the Eastern United States. Ecological Monographs. 68(4): 465–485.

Iverson, L.; Prasad, A.; Matthews, S.N. 2008a. Modeling potential climate change impacts on the trees of the Northeastern United States. Mitigation and Adaptation Strategies for Global Change. 13: 517–540.

Iverson, L.R.; Prasad, A.M.; Matthews, S.N.; Peters, M. 2008b. Estimating potential habitat for 134 Eastern U.S. tree species under six climate scenarios. Forest Ecology and Management. 254: 390–406.

Iverson, L.R.; Schwartz, M.W.; Prasad, A.M. 2004. How fast and far might tree species migrate in the Eastern United States due to climate change? Global Ecology and Biogeography. 13: 209–219.

Jaynes, R.A. 1978. A hydrologic model of aspen-conifer succession in the Western United States. Res. Pap. INT-213. Ogden, UT: U.S. Department of Agriculture, Forest Service, Intermountain Forest and Range Experiment Station. 17 p.

Johnston, B.C. 2001. Multiple factors affect aspen regeneration on the Uncompahgre Plateau, west-central Colorado. In: Shepperd, W.D.; Binkley, D.; Bartos, D.L.; Stohlgren, T.J.; Eskew, L.G., comps. Sustaining aspen in western landscapes: symposium proceedings. Gen. Tech. Rep. RMRS-P-18. Fort Collins, CO: U.S. Department of Agriculture, Forest Service, Rocky Mountain Research Station: 395–414.

Jones, J.R.; DeByle, N.V. 1985. Climates. In: DeByle, N.V.; Winokur, R.P., eds. Aspen: ecology and management in the Western United States. Gen. Tech. Rep. RM-119. Fort Collins, CO: U.S. Department of Agriculture, Forest Service, Rocky Mountain Forest and Range Experiment Station: 57–64.

Jones, J.R.; DeByle, N.V.; Bowers, D.M. 1985a. Insects and other invertebrates. In: DeByle, N.V.; Winokur, R.P., eds. Aspen: ecology and management in the Western United States. Gen. Tech. Rep. RM-119. Fort Collins, CO: U.S. Department of Agriculture, Forest Service, Rocky Mountain Forest and Range Experiment Station: 107–114.

Jones, J.R.; Kaufmann, M.R.; Richardson, E.A. 1985b. Effects of water and temperature. In: DeByle, N.V.; Winokur, R.P., eds. Aspen: ecology and management in the Western United States. Gen. Tech. Rep. RM-119. Fort Collins, CO: U.S. Department of Agriculture, Forest Service, Rocky Mountain Forest and Range Experiment Station: 71–76.

Jones, J.R. 1985. Distribution. In: DeByle, N.V.; Winokur, R.P., eds. Aspen: ecology and management in the Western United States. Gen. Tech. Rep. RM-119. Fort Collins, CO: U.S. Department of Agriculture, Forest Service, Rocky Mountain Forest and Range Experiment Station: 9–10.

Kay, C.E. 1997. Is aspen doomed? Journal of Forestry. 95: 4–11.

Kinney, K.K.; Lindroth, R.L.; Jung, S.M.; Nordheim, E.V. 1997. Effects of CO_2 and NO_3 availability on deciduous trees: phytochemistry and insect performance. Ecology. 78(1): 215–230.

Knowles, N.; Cayan, D. 2004. Elevational dependence of projected hydrologic changes in the San Francisco estuary and watershed. Climatic Change. 62: 319–336.

Krawchuk, M.A.; Cumming, S.G.; Flannigan, M.D.; Wein, R.W. 2006. Biotic and abiotic regulation of lightning fire initiation in the mixed-wood boreal forest. Ecology. 87(2): 458–468.

Kubiske, M.E.; Quinn, V.S.; Heilman, W.E.; McDonald, E.P.; Marquardt, P.E.; Teclaw, R.M.; Friend, A.L.; Karnosky, D.F. 2006. Interannual climatic variation mediates elevated CO^2 and O^3 effects on forest growth. Global Change Biology. 12: 1054–1068.

Kulakowski, D.; Veblen, T.T.; Drinkwater, S. 2004. The persistence of quaking aspen (*Populus tremuloides*) in the Grand Mesa area, Colorado. Ecological Applications. 14(5): 1603–1614.

Kulakowski, D.; Veblen, T.T.; Kurzel, B.P. 2006. Influences of infrequent fire, elevation and pre-fire vegetation on the persistence of quaking aspen (*Populus tremuloides* Michx.) in the Flat Tops area, Colorado, USA. Journal of Biogeography. 33: 1397–1413.

Kurzel, B.P.; Veblen, T.T.; Kulakowski, D. 2007. A typology of stand structure and dynamics of quaking aspen in northwestern Colorado. Forest Ecology and Management. 252: 176–190.

LaMalfa, E.M.; Ryel, R. 2008. Differential snowpack accumulation and water dynamics in aspen and conifer communities: implications for water yield and ecosystem function. Ecosystems. 11: 569–581.

Landhäusser, S.M.; Lieffers, V.J. 1998. Growth of *Populus tremuloides* in association with *Calamagrostis canadensis*. Canadian Journal of Forest Research. 28: 396–401.

Larsen, E.J.; Ripple, W.J. 2003. Aspen age structure in the northern Yellowstone ecosystem: USA. Forest Ecology and Management. 179: 469–482.

Lawrence, W.T.; Oechel, W.C. 1983. Effects of soil temperature on the carbon exchange of taiga seedlings. II. Photosynthesis, respiration, and conductance. Canadian Journal of Forest Research. 13: 850–859.

Lieffers, V.J.; Landhäusser, S.M.; Hogg, E.H. 2001. Is the wide distribution of aspen a result of its stress tolerance? In: Shepperd, W.D.; Binkley, D.; Bartos, D.L.; Stohlgren, T.J.; Eskew, L.G., comps. Sustaining aspen in western landscapes: symposium proceedings. Gen. Tech. Rep. RMRS-P-18. Fort Collins, CO: U.S. Department of Agriculture, Forest Service, Rocky Mountain Research Station: 311–324.

Lindroth, R.L. 2008. Chemical ecology of aspen: herbivory and ecosystem consequences. In: Restoring the West 2008: frontiers in aspen restoration. Logan, UT. http://extension.usu.edu/forestry/utahforests/rtw2006/RTW2006.htm. (15 March 2010).

Lindroth, R.L.; Kopper, B.J.; Parsons, W.F.J.; Bockheim, J.G.; Karnosky, D.F.; Hendrey, G.R.; Pregitzer, K.S.; Isebrands, J.G.; Sober, J. 2001. Consequences of elevated carbon dioxide and ozone for foliar chemical composition and dynamics in trembling aspen (*Populus tremuloides*) and paper birch (*Betula papyrifera*). Environmental Pollution. 115: 395–404.

Little, E.L., Jr. 1971. Atlas of United States trees. Volume 1: Conifers and important hardwoods. Misc. Publ. 1146. Washington, DC: U.S. Department of Agriculture, Forest Service.

Logan, J. 2008. Gypsy moth risk assessment in the face of a changing environment: a case history application in Utah and the greater Yellowstone ecosystem. In: Restoring the West 2008: frontiers in aspen restoration. Logan, UT. http://extension.usu.edu/forestry/utahforests/rtw2006/RTW2006.htm. (15 March 2010).

Manier, D.J.; Laven, R.D. 2001. Changes in landscape patterns and associated forest succession on the western slope of the Rocky Mountains, Colorado. In: Shepperd, W.D.; Binkley, D.; Bartos, D.L.; Stohlgren, T.J.; Eskew, L.G., comps. Sustaining aspen in western landscapes: symposium proceedings. Gen. Tech. Rep. RMRS-P-18. Fort Collins, CO: U.S. Department of Agriculture, Forest Service, Rocky Mountain Research Station: 15–26.

Martin, T.E. 2007. Climate correlates of 20 years of trophic changes in a high-elevation riparian system. Ecology. 88(2): 367–380.

Mastrandrea, M.D.; Tebaldi, C.; Snyder, C.P.; Schneider, S.H. 2009. Current and future impacts of extreme events. California Climate Change Center Final Report, CEC-500-2009-026-F: 81 p. http://www.energy.ca.gov/2009publications/ CEC-500-2009-026/CEC-500-2009-026-F.PDF. (15 March 2010).

Millar, C.I.; Woolfendon, W.B. 1999. The role of climate change in interpreting historical variability. Ecological Applications. 9(4): 1207–1216.

Moser, S.; Franco, G.; Pittiglio, S.; Chou, W.; Cayan, D. 2009. The future is now: an update on climate change science impacts and response options for California. California Climate Change Center Final Report, CEC-500-2008-071: 114 p. http://www.energy.ca.gov/2008publications/CEC-500-2008-071/CEC-500-2008-071.PDF. (15 March 2010).

Mueggler, W.F. 1985. Vegetation associations. In: DeByle, N.V.; Winokur, R.P., eds. Aspen: ecology and management in the Western United States. Gen. Tech. Rep. RM-119. Fort Collins, CO: U.S. Department of Agriculture, Forest Service, Rocky Mountain Forest and Range Experiment Station: 45–55.

Mueggler, W.F. 1988. Aspen community types of the Intermountain Region. Gen. Tech. Rep. INT-250. Ogden, UT: U.S. Department of Agriculture, Forest Service, Intermountain Research Station. 135 p.

Niinemets, U.; Valladares, F. 2006. Tolerance to shade, drought and waterlogging of temperate, Northern Hemisphere trees and shrubs. Ecological Monographs. 76: 521–547.

Nitschke, C.R.; Innes, J.L. 2008. A tree and climate assessment tool for modelling ecosystem response to climate change. Ecological Modelling. 210: 263–277.

Norby, R.J.; DeLucia, E.H.; Gielen, B.; Calfapietra, C.; Giardina, C.P.; King, J.S.; Ledford, J.; McCarthy, H.R.; Moore, D.J.P.; Ceulemans, R.; De Angelis, P.; Finzi, A.C.; Karnosky, D.F.; Kubiske, M.E.; Lukac, M.; Pregitzer, K.S.; Scarascia-Mugnozza, G.E.; Schlesinger, W.H.; Oren, R. 2005. Forest response to elevated CO^2 is conserved across a broad range of productivity. Proceedings of the National Academy of Sciences of the United States of America. 102(50): 18052–18056.

Peet, R.K. 2000. Forests of the Rocky Mountains. In: Barbour, M.G.; Billings, W.D., eds. North American terrestrial vegetation. New York: Cambridge University Press: 63–102.

Perala, D.A. 1990. *Populus tremuloides*. In: Burns, R.M.; Honkala, B.H., tech. eds. Silvics of North America: 2. Hardwoods. Agric. Handb. 654. Washington, DC: U.S. Department of Agriculture, Forest Service: 555–569.

Peterson, E.B.; Peterson, N.M. 1992. Ecology, management, and use of aspen and balsam poplar in the prairie provinces, Canada. Special Report 1. Edmonton, AB: Forestry Canada, Northwest Region, Northern Forestry Centre. 252 p.

Potter, D.A. 1998. Forested communities of the upper montane in the central and southern Sierra Nevada. Gen. Tech. Rep. PSW-169. Albany, CA: U.S. Department of Agriculture, Forest Service, Pacific Southwest Research Station. 319 p.

Pregitzer, K.S.; Zak, D.R.; Maziasz, J.; DeForest, J.; Curtis, P.S.; Lussenhop, J. 2000. Interactive effects of atmospheric CO_2 and soil-N availability on fine roots of *Populus tremuloides*. Ecological Applications. 10(1): 18–33.

Rehfeldt, G.E.; Ferguson, D.E.; Crookston, N.L. 2009. Aspen, climate, and sudden decline in western USA. Forest Ecology and Management. 258: 2353–2364.

Reich, P.B.; Bakken, P.; Carlson, D.; Frelich, L.E.; Friedman, S.K.; Grigal, D.F. 2001. Influence of logging, fire, and forest type on biodiversity and productivity in southern boreal forests. Ecology. 82(10): 2731–2748.

Richardson, T.W.; Heath, S.K. 2004. Effects of conifers on aspen-breeding bird communities in the Sierra Nevada. Transactions of the Western Section of the Wildlife Society. 40: 68–81.

Rodebaugh, D. 2008. The aspen conundrum. Durango Herald. http://archive. durangoherald.com/asp-bin/printable_article_generation.asp?article_path=/ earth/08/earth080221_1.htm. (27 November 2009).

Roden, J.S.; Pearcy, R.W. 1993. The effect of flutter on the temperature of poplar leaves and its implication for carbon gain. Plant, Cell & Environment. 16: 571–577.

Rogers, P. 2002. Using forest health monitoring to assess aspen forest cover change in the southern Rockies ecoregion. Forest Ecology and Management. 155: 223–236.

Romme, W.H.; Floyd-Hanna, L.; Hanna, D.D.; Bartlett, E. 2001. Aspen's ecological role in the West. In: Shepperd, W.D.; Binkley, D.; Bartos, D.L.; Stohlgren, T.J.; Eskew, L.G., comps. Sustaining aspen in western landscapes: symposium proceedings. Gen. Tech. Rep. RMRS-P-18. Fort Collins, CO: U.S. Department of Agriculture, Forest Service, Rocky Mountain Research Station: 243–260.

Rood, S.B.; Patino, S.; Coombs, K.; Tyree, M.T. 2000. Branch sacrifice: cavitation-associated drought adaptation of riparian cottonwoods. Trees. 14: 248–257.

Ryel, R.; Bartos, D. 2008. Aspen classification: the need for an effective functional system of classification. In: Restoring the West 2008: frontiers in aspen restoration. Logan, UT. http://extension.usu.edu/forestry/utahforests/rtw2006/RTW2006.htm. (15 March 2010).

Sawyer, J.O.; Keeler-Wolf, T. 1995. A manual of California vegetation. Sacramento, CA: California Native Plant Society. 471 p.

Schier, G.A.; Shepperd, W.D.; Jones, J.R. 1985. Regeneration. In: DeByle, N.V.; Winokur, R.P., eds. Aspen: ecology and management in the Western United States. Gen. Tech. Rep. RM-119. Fort Collins, CO: U.S. Department of Agriculture, Forest Service, Rocky Mountain Forest and Range Experiment Station: 197–208.

Sexton, J.O.; Ramsey, R.D.; Bartos, D.L. 2006. Habitone analysis of quaking aspen in the Utah Book Cliffs: effects of site water demand and conifer cover. Ecological Modelling. 198: 301–311.

Shepperd, W.D. 2008. Sudden aspen decline in the Western U.S.: introduction and background [Poster]. In: Sudden aspen decline (SAD) meeting. Fort Collins, CO. http://www.aspensite.org/. (15 March 2010).

Shepperd, W.D.; Guyon, J. 2006. Massive aspen die-off in the Western U.S.: What is going on? In: Restoring the West 2008: frontiers in aspen restoration. Logan, UT. http://extension.usu.edu/forestry/utahforests/rtw2006/RTW2006.htm. (15 March 2010).

Shepperd, W.D.; Rogers, P.C.; Burton, D.; Bartos, D.L. 2006. Ecology, biodiversity, management, and restoration of aspen in the Sierra Nevada. Gen. Tech. Rep. GTR-RMRS-178. Fort Collins, CO: U.S. Department of Agriculture, Forest Service, Rocky Mountain Research Station. 122 p.

Shields, W.J.; Bockheim, J.G. 1981. Deterioration of trembling aspen clones in the Great Lakes region. Canadian Journal of Forest Research. 11: 530–537.

Smith, A.E.; Smith, F.W. 2005. Twenty-year change in aspen dominance in pure aspen and mixed aspen/conifer stands on the Uncompahgre Plateau, Colorado, USA. Forest Ecology and Management. 213: 338–348.

Spracklen, D.V.; Mickley, L.J.; Logan, J.A.; Hudman, R.C.; Yevich, R.; Flannigan, M.D.; Westerling, A.L. 2009. Impacts of climate change from 2000 to 2050 on wildfire activity and carbonaceous aerosol concentrations in the Western United States. Journal of Geophysical Research. 114: D20301.

St. Clair, S. 2008. Physiological ecology of aspen-conifer interactions. In: Restoring the West 2008: frontiers in aspen restoration. Logan, UT. http://extension.usu.edu/forestry/utahforests/rtw2006/RTW2006.htm. (15 March 2010).

Stohlgren, T.J.; Binkley, D.; Chong, G.W.; Kalkhan, M.A.; Schell, L.D.; Bull, K.A.; Otsuki, Y.; Newman, G.; Bashkin, M.; Son, Y. 1999. Exotic plant species invade hot spots of native plant diversity. Ecological Monographs. 69: 25–46.

Turner, M.G.; Romme, W.H.; Reed, R.A.; Tuskan, G.A. 2003. Post-fire aspen seedling recruitment across the Yellowstone (USA) landscape. Landscape Ecology. 18: 127–140.

Tuskan, G.A.; Francis, K.E.; Russ, S.L.; Romme, W.H.; Turner, M.G. 1996. RAPD markers reveal diversity within and among clonal and seedling stands of aspen in Yellowstone National Park, U.S.A. Canadian Journal of Forest Research. 26: 2088–2098.

Van Wagner, C.E. 1977. Conditions for the start and spread of crown fire. Canadian Journal of Forest Research. 7: 23–34.

Wan, X.; Zwiazek, J.J. 1999. Mercuric chloride effects on root water transport in aspen seedlings. Plant Physiology. 121: 1–8.

Weixelman, D.A.; Zamudio, D.C.; Zamudio, K.A. 1999. Eastern Sierra Nevada riparian field guide. R4-ECOL-99-01. Ogden, UT: U.S. Department of Agriculture, Forest Service.

Westerling, A.L.; Hidalgo, H.G.; Cayan, D.R.; Swetnam, T.W. 2006. Warming and earlier spring increase Western US forest wildfire activity. Science. 313(5789): 940–943.

Wirth, T.; Maus, P.; Powell, J.; Lachowski, H. 1996. Monitoring aspen decline using remote sensing and GIS: Gravelly Mountain Landscape, Southwestern Montana. Prepared for the Remote Sensing Committee, U.S. Department of Agriculture, Forest Service, Salt Lake City, UT.

Worrall, J.J.; Egeland, L.; Eager, T.; Mask, R.A.; Johnson, E.W.; Kemp, P.A.; Shepperd, W.D. 2008. Rapid mortality of *Populus tremuloides* in southwestern Colorado, USA. Forest Ecology and Management. 255: 686–696.

Zoltai, S.C.; Singh, T.; Apps, M.J. 1991. Aspen in a changing climate. In: Navratil, S.; Chapman, P.B., eds. Aspen management for the 21st century, proceedings symposium. Edmonton, AB: Forestry Canada Northwest Region and Poplar Council of Canada: 143–152.

Appendix:
Introduction of a Sudden Aspen Decline (SAD) Survey Tool (developed by T.L. Morelli)

Toni Lyn Morelli
Pacific Southwest Research Station
USDA Forest Service
morellitlm@gmail.com

David Burton
Aspen Delineation Project
peregrines@prodigy.net

The purpose of this survey is to record baseline data for aspen stands in areas where sudden aspen decline (SAD) is not occurring, as well as to identify any stands where SAD is occurring.

The SAD Survey is designed to recognize potential symptoms of SAD as well as symptoms of general aspen decline and excessive browsing pressure. It is deliberately short (just 1 page printed double-sided) and does not require conducting transects. Moreover, the SAD survey does not require expertise in botany or forestry. If there are questions that go beyond your expertise, leave these sections blank.

The SAD survey is in a preliminary stage. Any federal employee interested in using the survey, or making revisions to it, can contact one of the coordinators above; we'd be happy to hear ideas.

Instructions for conducting survey—
This survey should be conducted at the center of the aspen stand. It is best to conduct the survey after leaves have flushed and before they have fallen (late spring to early fall, depending on your location).

Answer the questions from what you know of the stand. Thus, if the stand is small enough that you are able to see/walk through the whole stand, answer the questions about the entire stand. If the stand is large, answer the questions about what you can see from the center of the stand. Question 12 addresses the issue of whether the edge of the stand is different from the center.

Your contact information will be useful in case there is anything on the sheet that we cannot read or is otherwise unclear.

Use the Stand ID that makes sense for your federal agency. U.S. Forest Service should use a Forest Code and District Code, plus the Stand Code when possible.

Check the box or boxes for primary stand type that seems most fitting. If there is another stand type that you think would be more appropriate (e.g., snowpocket), you can write it in, but please also check at least one of the four present boxes.

Primary stand type is defined as follows (Shepperd et al. 2006): Mark "slope" if the aspen stand is found on a hill. Mark "lithic" if the aspen are found in or next to a talus or other rocky field. Mark "meadow fringe" if the aspen stand occurs on the edge of a meadow. Mark "riparian" if the aspen stand occurs in an area with permanent or seasonal standing or moving water.

"Primary aspen form" is to distinguish the more common tree form from the shrub-like krummholz form.

Questions:

1. Estimate the percentage of the canopy that shows recent crown loss, which is thinning of the foliage or branch dieback. The categories are 0–33%, 34–66%, and 66–100%.

2. Estimate the percentage of the canopy that has died and fallen. The categories are 0–33%, 34–66%, and 67–100%.

3. Estimate the percentage of the stand that has died and is still standing. Then estimate the percentage of those dead and still standing trees that died recently. These trees can be distinguished by their white color as they will still have most of their bark. The categories are 0–33%, 34–66%, and 67–100%.

4. Estimate the size class for the majority (greater than 50%) of aspen that are standing and alive. The categories are less than 1 in (2.54 cm) dbh, 1–8 in (2.54–20.32 cm) dbh, and greater than 8 in (20.32 cm) dbh.

5. Estimate the size class for the majority (greater than 50%) of mortality that has occurred recently (indicated by their white color as they will still have most of their bark). The categories are less than 1 in (2.54 cm) diameter at breast height (dbh), 1–8 in (2.54–20.32 cm) dbh, and greater than 8 in (20.32 cm) dbh.

6. Compare the absolute number of young aspen of 1–5 in (2.54–12.7 cm) dbh in the stand with the absolute number of mature aspen trees (greater than 8 in/20.32 cm). For example, if you estimate 60 young aspen and 10 large mature aspen, approximate your answer by checking the "5x" box. If you can see 2 young aspen and 17 mature aspen, check the "<1x" box.

7. Compare the number of sprouts (less than 1 in/2.54 cm dbh) in the stand with the number of mature aspen trees (greater than 8 in/20.32 cm). For example, if you can see approximately 500 sprouts and 30 large aspen stems, answer "10x". If you can see no or very few sprouts, answer "<1x".

8. Estimate the percentage of sprouts (less than 1 in dbh) that show evidence of herbivory. Evidence includes any chewed leaves or stems. Also note if the herbivory appears light, moderate, or heavy. Thus, in an area with consistent

but low-density deer use, you might see one chewed leaf from 80% of sprouts. In this case, you should mark "76–99%" and "light."

9. Describe the pathogen or insect damage that you see, including the species if you can identify them. Note any beaver damage observed.
10. Note any evidence of fire in the stand.
11. Note the presence of conifers, which section and portion they make up of the forest, and which species predominates, if you know.
12. Now walk to the edge of the stand and note whether sprouting is uniform across the stand. For example, in some stands there are more sprouts at one edge than other edges or in the center. Also note if there are patches of mortality in different parts of the stand.

Finally, there is a section to make notes if you run out of room on the form or have anything to add.

If you can take pictures, they will help supplement this information, and will be much appreciated.

Comments, questions, pictures, or completed surveys?

Contact David Burton at 916-663-2574, peregrines@prodigy.net
Aspen Delineation Project, P.O. Box 348, Penryn, CA 95663

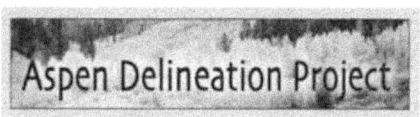

Sudden Aspen Decline (SAD) Survey

CONDUCT THIS SURVEY IN CENTER OF STAND DURING SPRING OR SUMMER

DATE:_____

Surveyor's Name: _____   Surveyor's Phone #_____

Stand Center GPS Coordinates: _____ N_____ W units:_____

Stand ID: _____   Stand size: _____ ha

Environmental context: Elevation:_____m   Aspect:_____°   Slope:_____%

Primary stand type: Slope ☐   Lithic ☐   Meadow Fringe ☐   Riparian ☐

Primary aspen form: Upright tree ☐   Shrub ☐

1. How much recent crown loss (thinning of the foliage and/or branch dieback) across the canopy?
    <34% ☐   34–66% ☐   >66% ☐

2. What percentage of the stand is down and dead aspen?
    <34% ☐   34–66% ☐   >66% ☐

3. What percentage of the stand is standing and dead aspen?
    0–25% ☐   26–50% ☐   51–75% ☐   76–100% ☐
    Do they appear recently dead (bark still intact)?
                <34% ☐   34–66% ☐   >66% ☐

4. The Majority (>50%) of live aspen is in which size class?
    <1 inch dbh ☐   1–8 inch dbh ☐   >8 inch dbh ☐

5. The majority of current aspen mortality is located in which size classes?
    <1 inch dbh ☐   1–8 inch dbh ☐   >8 inch dbh ☐

6. How many more young established aspen (1–5 inch dish) are present than mature aspen (>8 inches)?
    <1x ☐   2x ☐   5x ☐   >10x ☐

7. How many more aspen sprouts (<1 inch dbh) are present than mature aspen (>8 inches)?
    <1x ☐   10x ☐   100x ☐   1000x ☐

8. What percentage of sprouts (<1 inch dbh) show any evidence of ungulate (e.g., elk, deer, cattle, sheep) herbivory?
    0–25% ☐   26–50% ☐   51–75% ☐   76–100% ☐
    Is the sprout herbivory   light ☐   moderate ☐   heavy ☐

9. Is there evidence of pathogens or insect damage? Is there evidence of beaver damage? Describe

_____

_____

_____

10. Is there any evidence of past fire in the stand?
    Yes ☐   No ☐

11. What is the size of conifers within this stand?
    no conifers ☐   conifer in understory ☐   mixed in canopy ☐   majority of canopy ☐
    Primarily which species of conifer? _____

12. Walk to the edge of the stand. Is sprouting uniform across the stand?
    Yes ☐   No ☐   Explain: _____
    Is morality uniform across the stand?   _____
    Yes ☐   No ☐   _____

Sudden Aspen Decline (SAD) Survey (continued)

GENERAL NOTES: _____

_____

_____

_____

_____

_____

_____

_____

_____

_____

If possible, please take at least three pictures: the environmental context (broader view), a stand shot, and, where appropriate, a closeup showing environmental damage.

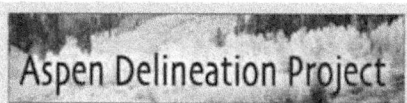

Comments, questions, or completed surveys/pictures?
Contact David Burton at 916-663-2574, peregrines@prodigy.net
Aspen Delineation Project, P.O. Box 348, Penryn, CA 95663

Developed by Toni Lyn Morelli, Pacific Southwest Research Station, and David Burton, Aspen Delineation Project.
Thanks to John Guyon, Ted Hogg, Connie Millar, Paul Rogers, Wayne Shepperd, and Jim Worrall for comments.

Pacific Northwest Research Station

Web site	http://www.fs.fed.us/pnw
Telephone	(503) 808-2592
Publication requests	(503) 808-2138
FAX	(503) 808-2130
E-mail	pnw_pnwpubs@fs.fed.us
Mailing address	Publications Distribution
	Pacific Northwest Research Station
	P.O. Box 3890
	Portland, OR 97208-3890

www.ingramcontent.com/pod-product-compliance
Lightning Source LLC
Chambersburg PA
CBHW050632210526
45168CB00025BA/3398